AF269779

AREA 51

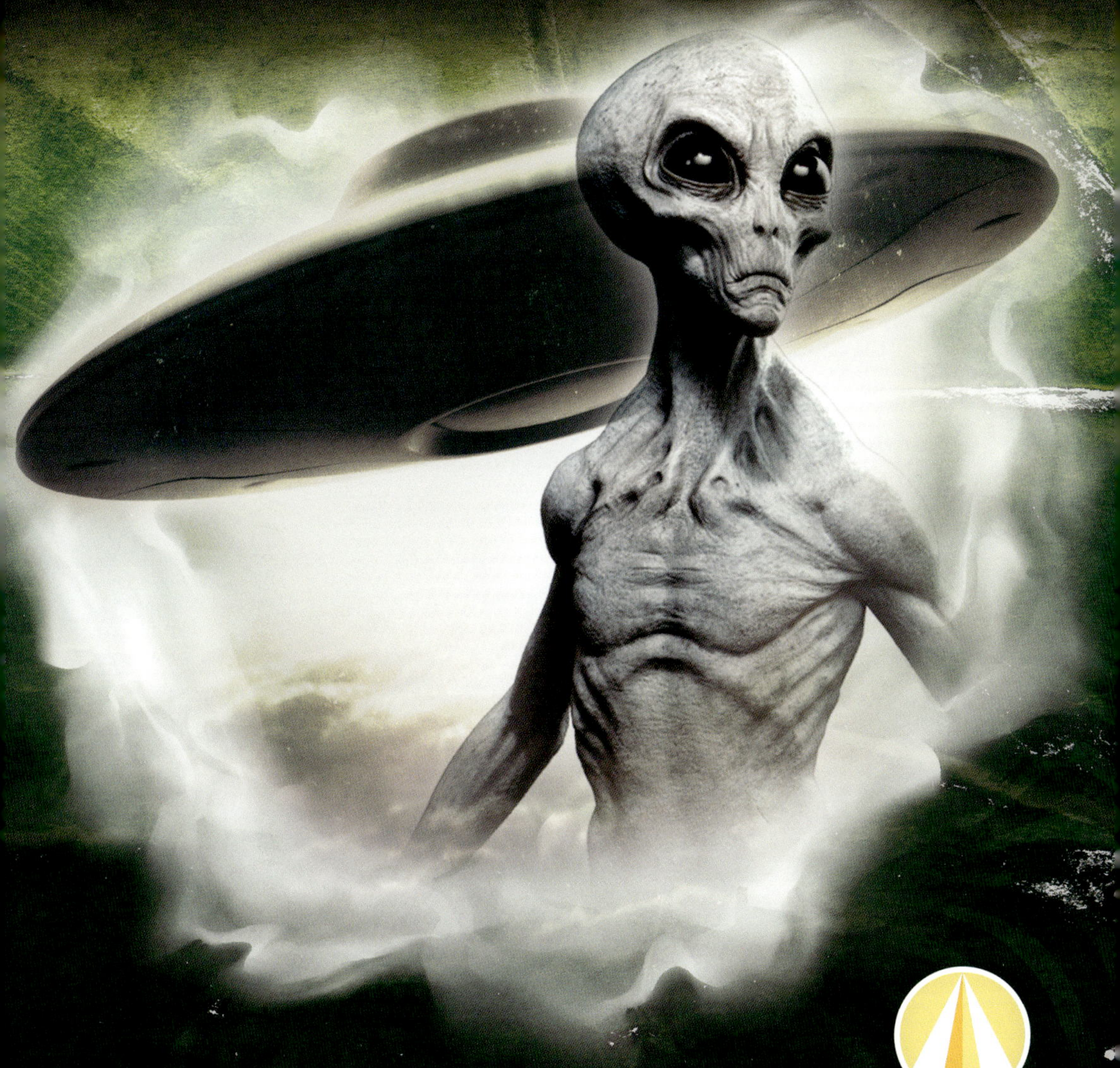

An Imprint of Abdo Zoom
abdobooks.com

KENNY ABDO

abdobooks.com

Published by Abdo Zoom, a division of ABDO, P.O. Box 398166, Minneapolis, Minnesota 55439. Copyright © 2020 by Abdo Consulting Group, Inc. International copyrights reserved in all countries. No part of this book may be reproduced in any form without written permission from the publisher. Fly!™ is a trademark and logo of Abdo Zoom.

Printed in the United States of America, North Mankato, Minnesota.
102019
012020

Photo Credits: Alamy, iStock, Shutterstock
Production Contributors: Kenny Abdo, Jennie Forsberg, Grace Hansen
Design Contributors: Dorothy Toth, Neil Klinepier, Pakou Moua

Library of Congress Control Number: 2019941603

Publisher's Cataloging-in-Publication Data

Names: Abdo, Kenny, author.
Title: Area 51 / by Kenny Abdo
Description: Minneapolis, Minnesota : Abdo Zoom, 2020 | Series: Guidebooks to the unexplained | Includes online resources and index.
Identifiers: ISBN 9781532129339 (lib. bdg.) | ISBN 9781644942864 (pbk.) | ISBN 9781098220310 (ebook) | ISBN 9781098220808 (Read-to-Me ebook)
Subjects: LCSH: Area 51 (Nev.)--Juvenile literature. | Extraterrestrial beings--Juvenile literature. | Conspiracy--Juvenile literature. | Legends--Juvenile literature. | UFOs--Juvenile literature.
Classification: DDC 001.942--dc23

TABLE OF CONTENTS

AReA 51

WARNING
Restricted Area
It is unlawful to enter this area without permission of the Installation Commander.
Sec. 21, Internal Security Act of 1950; 50 U.S.C. 797
While on this installation all personnel and the property under their control are subject to search.
Use of deadly force authorized.
WARNING!
NO TRESPASSING
AUTHORITY N.R.S. 207-200
MAXIMUM PUNISHMENT: $1000 FINE
SIX MONTHS IMPRISEMENT
OR BOTH
STRICTLY ENFORCED
PHOTOGRAPHY
OF THIS AREA
IS PROHIBITED
18 U.S.C. 795

Few people have walked down the corridors of Area 51. Its highly sealed doors hold some of the best kept secrets in the history of the world.

People were told that Area 51 did not exist for many years. But today, more information about the top-secret facility has come to light.

CLASSIFICATION

Area 51 is 80 miles (129 km) northwest of Las Vegas. It was built to test nuclear weapons in the 1950s.

Top-secret aircrafts were also tested there. Planes like the A-12 Blackbird and F-117 Nighthawk were worked on in Area 51 by the US government.

US officials have said that they had no interest in aliens or **UFOs**. However, some have linked Area 51 to an **alleged** government cover-up.

DECLASSIFIED

In 1947, farmer William Brazel found metallic rods, pieces of plastic, and silvery paper scraps scattered around his land in Roswell, New Mexico.

Roswell D[aily]

eased Wire — Associated Press

NUMBER 100   ESTABLISHED 1888   ROSWELL, NEW MEXICO,

Gen. Ramey Empt[ies]

Lewis Pushes Advantage in New Contract

Southern Mines Only Hold-outs In New Contract

Washington, July 9, (AP) — The odds lengthened today that John L. Lewis would play his new, ace-studded contract into a grand slam.

With 75 per cent of the soft coal industry signed up for work and shooting at full production by tomorrow, Southern operators still held out against the unprecedented wage pact signed yesterday by most Northern and Western producers.

The Southern Coal Producers association prepared to make its "final decision" at a noon meeting today. Its 100,000 workers are idle.

But one association member acknowledged privately that it looked as though, sooner or later all would be "forced" to accede. Lewis, it was learned, rejected their request to alter some of the terms in a 90-minute session yesterday.

Federal labor officials conceded it would be difficult for the South to hold out alone, with the rest of the country producing and selling coal—at a price perhaps 70 cents to $1 a ton higher than be-

...proclaimed his own certain outcome.

..."reasonable to assume," he ...telling reporters of the United Mine Workers' fat contract ...that the rest of the

Sheriff Wilcox Takes Leading Role in Excitement Over Report 'Saucer' Found

That worried look on the face of Sheriff George Wilcox, in the picture above, comes from having been cast, more suddenly than he liked, into the role of leading man in the Roswell saucer...

U. S. Lend-Lease Mill Looms

Arrest 2,000 In Athens in Commie Plot

Revolution Was Set to Be Pulled Off Thursday

Athens, July 9 (AP)—The Greek government announced that more than 2,000 persons were arrested in the Athens area early today in raids aimed at stamping out a Communist plot to stage a revolution and spread civil war throughout the country.

Minister of Public Order Napoleon Zervas said the zero hour for the Communist stroke was to have been around 1 a. m. tomorrow, when attacks were to have been staged simultaneously in parts of Greece, bringing the present mountain guerilla warfare into urban centers.

Between 3,000 and 4,000 police, gendarmes and soldiers staged the lightning raids before dawn this morning, Zervas said. He added that many important Communists already had fled and either were hiding in Athens or in the mountains.

Most of those arrested, he said, will be taken to islands near Athens, while the investigation continues.

The transport already has begun. Some ringleaders, Zervas added, will remain in Athens to await hearings. Those not implicated in the plot will be released and others probably will be exiled, officials said.

A leftist leader who escaped arrest in the first raids declared: "They're making a clean sweep"

—0—

Brazel reported the strange objects to the local sheriff. The military arrived shortly after and carted the debris away.

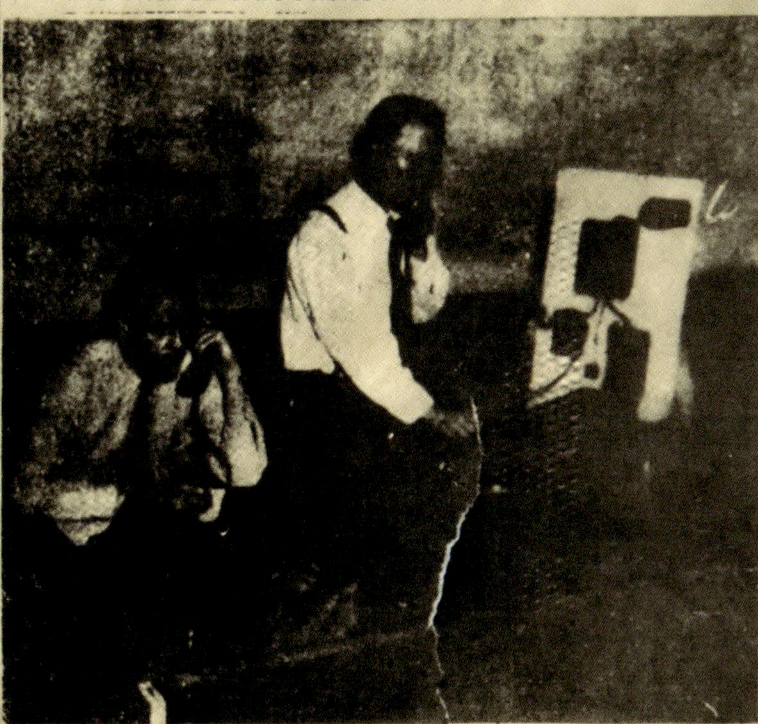

Area 51 is believed to be where aliens and their spacecraft are kept.

Other locals have reported seeing strange lights around Area 51 for many years. People believe these lights are an alien aircraft taking off or being tested in the desert.

Ray Santilli released a video in 1995. It showed an **alleged** alien **autopsy** after the Roswell crash. This added to the alien cover-up **theories**.

In 2017, the **Pentagon** said that there was a new government program. It was created to study irregular aerospace threats. Some people think this refers to **UFOs**.

AREA 51
ALIEN
Alien Research Center

IN MEDIA

Area 51 **conspiracies** go beyond aliens and **UFOs**. Other **theories** about the facility have popped up in the news over the years. Like the creation of time travel and weather control inside of Area 51!

Even though Area 51 is the setting for many hit movies, do not try to approach it. You just might become an alien's new cellmate, or worse!

GLOSSARY

allege – to claim something was done without proof.

autopsy – to study something after it has died.

conspiracy – an unproven theory stated as fact.

debris – scattered pieces of rubble.

Pentagon – the headquarters for the United States Department of Defense.

theory – an idea used to justify something that happened.

UFO – a mysterious object seen in the sky that has no explanation. UFO is short for Unidentified Flying Object.

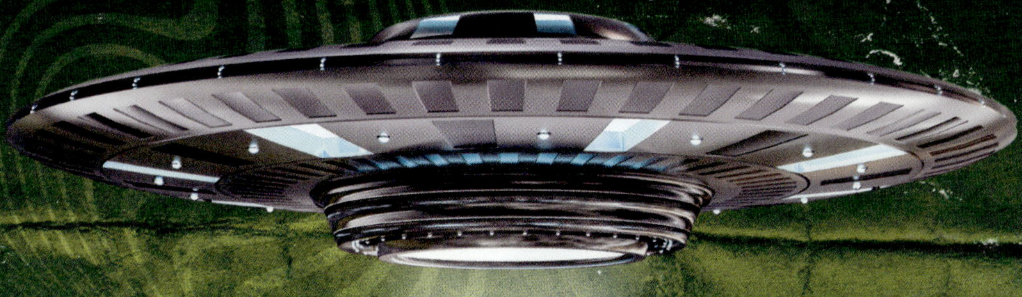

ONLINE RESOURCES

To learn more about Area 51, please visit **abdobooklinks.com** or scan this QR code. These links are routinely monitored and updated to provide the most current information available.

INDEX